T0224995

SpringerBriefs in Architectural Design and Technology

Series Editor

Thomas Schröpfer, Architecture and Sustainable Design, Singapore University of Technology and Design, Singapore, Singapore

Indexed by SCOPUS

Understanding the complex relationship between design and technology is increasingly critical to the field of Architecture. The *Springer Briefs in Architectural Design and Technology* series provides accessible and comprehensive guides for all aspects of current architectural design relating to advances in technology including material science, material technology, structure and form, environmental strategies, building performance and energy, computer simulation and modeling, digital fabrication, and advanced building processes. The series features leading international experts from academia and practice who provide in-depth knowledge on all aspects of integrating architectural design with technical and environmental building solutions towards the challenges of a better world. Provocative and inspirational, each volume in the Series aims to stimulate theoretical and creative advances and question the outcome of technical innovations as well as the far-reaching social, cultural, and environmental challenges that present themselves to architectural design today. Each brief asks why things are as they are, traces the latest trends and provides penetrating, insightful and in-depth views of current topics of architectural design. *Springer Briefs in Architectural Design and Technology* provides must-have, cutting-edge content that becomes an essential reference for academics, practitioners, and students of Architecture worldwide.

Joerg Baumeister · Ioana C. Giurgiu

SeaOasis

Floating Aquaculture for Smallholders'
Global Food Security

 Springer

Joerg Baumeister
Gold Coast Campus
SeaCities, Griffith University
Southport, QLD, Australia

Ioana C. Giurgiu
Gold Coast Campus
SeaCities, Griffith University
Southport, QLD, Australia

ISSN 2199-580X   ISSN 2199-5818 (electronic)
SpringerBriefs in Architectural Design and Technology
ISBN 978-981-19-1372-3   ISBN 978-981-19-1373-0 (eBook)
https://doi.org/10.1007/978-981-19-1373-0

This Springer imprint is published by the registered company Springer Nature Singapore Pte Ltd.
The registered company address is: 152 Beach Road, #21-01/04 Gateway East, Singapore 189721, Singapore

Introduction

After thousands of years of evolution, humankind has achieved a stage of civilization where many people take it for granted to enjoy strawberries in the wintertime, drink water from the other side of the globe, and fly around the world (if effective vaccinations against threatening viruses are available).

Unfortunately, this quality of life is often achieved at the expense of others suffering from starvation, water shortage, or environmental pollution. This fatal unbalance will likely increase due to a constantly growing population.

The problem of ensuring food security and production at sufficient levels to support future populations is among the key challenges for the 2050 horizon. Considering the combined effects of population growth, soil degradation due to climate change as well as limited available agricultural land, sustainable, and holistic approaches and design solutions must be explored.

The SeaOasis is an experimental design response aimed at tackling the challenge of increasing food security through sustainable design. Problems and opportunities relating to current food supply were analyzed to formulate a specific design brief. The brief aimed to identify key research directions that could support the improvement and sustainable future growth of current food systems through targeted design interventions and innovation.

The basic hypotheses of utilizing water surfaces to expand agricultural land, tailoring design solutions to smallhold farming and hybridizing traditional and industrial sustainable farming technologies derived from the brief were further applied to develop a practical design for a novel aqua-farming system: SeaOasis.

Appraised in terms of potential implementation locations and scale as well as economic feasibility, the SeaOasis provides a potent example for an economically viable, scalable modular solution, with potential applications in a variety of locations across the globe.

Contents

Chapter 1
Brief

1.1 Global Food Supply

Fifty years ago, only half as many people lived on the planet. Current studies predict a similar increase resulting in a population of nearly 10 billion (bn.) in the next 30 years (Fig. 1.1 a). Given the trend of rising food insecurity with ca. one in ten people in the world exposed to severe food insecurity levels in 2019 [1], it is likely that population growth will result in a proportional increase in the gap between food supply and demand.

If current yields and agricultural intensity remain constant and considering the potential impact of further expanding land-based agricultural areas (see options 1 and 2 below), it can be argued that global peak agricultural land (crop, grazing and feed culture lands) availability has already been reached.

Assuming all regions of the world would adopt average dietary choices (OWID 2011 world average [3]) over the next 30 years, food insecurity will likely grow dramatically due to a lack of 2 billion hectares (bn. ha.) agricultural land required to meet dietary demands (Fig. 1.1 b).

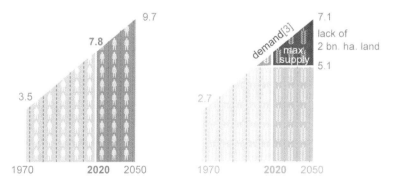

Fig.1.1 a World population growth (bn. people) [2], **b** availability of agricultural land (bn. ha.)

Conversely, if the entire world would adopt the United States' dietary choices, an additional 14 bn. ha. would be needed by 2050.

Looking at the earth's total surface, less than a third is covered by land. The 14,8 bn. ha. of land encompass areas which are not suitable for agriculture like barren land, glaciers, freshwater, urban land, and shrubs (in total 5,8 bn. ha.). The rest of the earth's surface is currently dedicated to forests (3,9 bn. ha.), livestock (4,0 bn. ha.), and crops (1,1 bn. ha.) (Fig. 1.2 a).

Logically, there are only a few options to accommodate the required extra 2,0 bn. ha. land for agriculture on appropriate land: Either by converting forests (option 1) or land for livestock (option 2) into agriculture.

Fig. 1.2 b demonstrates the extreme impact of using existing forested or grazing lands for additional crop areas. In option 1, half of the global forests would be destroyed. The loss of these forests would boost greenhouse gas (GHG) emissions, which would unbalance the global climatic system even further.

Option 2 (Fig. 1.2 b) re-purposes areas for livestock breeding into agricultural land. It seems to be more convenient than option 1, with the added benefit of reduced GHG production. However, to implement this option, dietary preferences, especially of rich nations, would have to change significantly from meat towards vegan nutrition which would be challenging to enforce.

Both options fail to identify sustainable solutions, thus, prompting the exploration of other opportunities for the expansion needed for future food production.

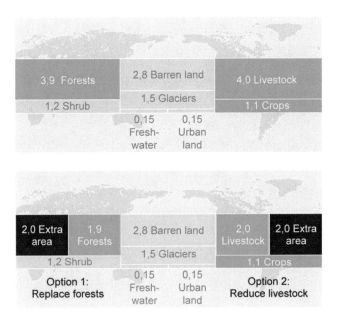

Fig.1.2 a Global land use for food production (bn. ha.) [4] , **b** options for additional agricultural land (bn. ha.)

If the usage of land areas for additional food production does not provide sustainable future solutions, water surfaces, covering more than two-thirds of the earth, represent a less explored alternative. Could the water surface provide a feasible option 3 by expanding food production to include and rely more on aquaculture (Fig. 1.3)?

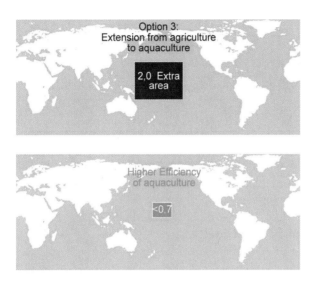

Fig. 1.3 a Option 3 for future food production (bn. ha.), **b** reduction due to higher efficiencies (bn. ha.)

Compared to common terrestrial crops such as wheat, pulse legumes and soybean yielding 1.1, 1-2 and 0.6-1.2 protein tons/ha/year respectively, viable alternative food sources such as microalgae cultures yield between 4 and 15 protein tons/ha/ year [5], thus dramatically reducing the area needed for future food crops.

Aquacultural food production is based on a long-standing tradition and has been growing by 7,5% per year since 1970 [6] (Fig. 1.4). The United Nations' Food and Agriculture Organization (FAO) highlight its crucial role in the global food supply with recent expansion of this sector reportedly leading to "improved nutrition and food security" [6].

On the other hand, due to the water's fluidity, aquaculture requires enclosures and other structures to support crop growth, which makes growing food in an aquatic environment more elaborate than on land. How far is it possible to counterbalance an increased effort of aqua-cultural food production with its higher efficiencies? And how far is this possible in a more sustainable way than current agriculture?

The research will start to answer these questions, exploring sustainable aquaculture principles, applying these principles to develop a design solution (chapter 2), and evaluating corresponding technical and economic implications (chapter 3). In conclusion, an indicative feasibility appraisal for the suggested expansion onto the sea is provided.

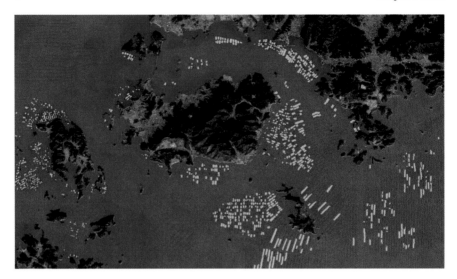

Fig. 1.4 Example of seaweed farmed on ropes in the shallow waters of South Korea, Sisan Island (Photo NASA [7])

1.2 Global Food Security

According to the definition adopted at the 1996 World Food Summit, "food security exists when all people, at all times, have physical and economic access to sufficient safe and nutritious food that meets their dietary needs and food preferences for an active and healthy life". [8]

The factors impacting access to adequate food sources, especially in developing areas of the world, are complex and cannot be traced back to a single cause. However, based on the above definition, the FAO identifies availability, access, stability, and utilization as the four key factors which, combined, determine the level of food security experienced [8].

Utilisation = Diverse diet of safe and nutritious food for a healthy life

Stability = Availability, access and utilization over time

Availability = Food production areas, sufficient yields, stocks and net trade

Access = Locally available and affordable food

Fig. 1.5 Four dimensions of food security (based on [8])

Therefore, to meet global food demand for 2050 and beyond, the aspects of location and accessibility, market prices, and affordability, as well as dietary requirements and preferences must be considered in relation to resource use and availability of agri- or aquacultural area over time.

In the context of globally meeting the UN Sustainable Development Goal (SDG) 2, which aims to "end hunger, achieve food security and improved nutrition and promote sustainable agriculture" [9], developing solutions that target the regions most affected by food insecurity (Fig. 1.6) is a key factor. At the same time, both agricultural and aquacultural area availability in these regions is greater than in the more developed regions of the world, providing a potential solution for meeting the additional demand for agricultural land in 2050.

GHI severity scale: ❧Alarming ❧Serious ❧Moderate

Fig. 1.6 Global hunger index 2020 [10]

1.3 The Role of Smallholders

SDG target 2.3 specifically addresses smallholders and, by 2030, aims to "double the agricultural productivity and incomes of small-scale food producers, in particular women, indigenous peoples, family farmers, pastoralists and fishers, including through secure and equal access to land, other productive resources and inputs, knowledge, financial services, markets and opportunities for value addition and nonfarm employment". [9]

While definitions relating to smallholder farms vary widely across countries, the term is generally used to refer to small-scale farms (often below 2 ha), a vast majority of which are family farms. Prompted by the UN recognition of the role of smallholders in meeting the sustainability goals, and designation of 2019-2028 as the decade of family farming [11], significant research has been undertaken to understand the challenges faced by small-scale farmers and how best to support the sustainable integration of small-scale farming into global markets.

At the global scale, smallholders are estimated to produce "28–31% of total crop production and 30–34% of food supply on 24% of gross agricultural area". [12] Given SDG target 2.3, small family farms will likely play an increasingly important part in global food supply chains.

The worldwide distribution of small-scale farms (Fig. 1.7) correlates with countries (especially in sub-Saharan Africa and Asia) with low and medium incomes, large numbers of rural communities, and high food insecurity levels. In these areas, smallholders play an essential part in terms of food supply and production and "creating employment in rural areas, reducing poverty and enhancing the sustainable management of natural resources".[13]

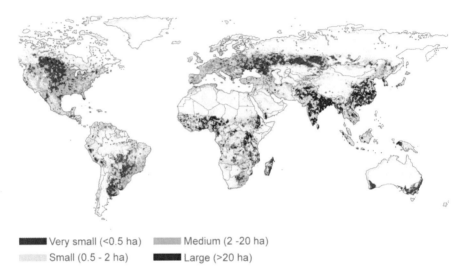

■■■■■ Very small (<0.5 ha)   ▓▓▓▓ Medium (2 -20 ha)
░░░░ Small (0.5 - 2 ha)   ■■■■ Large (>20 ha)

Fig. 1.7 Farm field sizes [14]

Especially for smallholder farmers or fishers, aquaculture is a "favorable livelihood source" which generates "substantial socioeconomic benefits to marginalized coastal communities in developing countries" [15].

Albeit their importance, smallholders face several challenges including limited land availability as well as limited access to international markets and technology which in turn affects their ability to produce high-value crops.

If tailored to smallholders, future aquatic food production solutions may provide the necessary additional crop areas for 2050, while at the same time supporting the economic growth of small-scale farms as well as linking local production to international markets by enhancing the ability to produce high-value crops via the integration of efficient and sustainable aquaculture technologies.

1.4 Closed-Loop Systems

The notion of closed-loop technology is used here to describe the available industrial technologies relating to sustainable and efficient food production, which are meant to function in a zero-waste, cyclical fashion, and apply re-use and recycling strategies to achieve this.

While traditional agricultural systems such as ancient oases successfully apply similar principles, high-tech industrial food production solutions provide promising alternatives with higher efficiencies in terms of yields, optimal land use, and overall system stability.

Historically, the Apollo mission and space race prompted the development of several technologies meant to sustain closed-loop food production in space (e.g. growing plants in water and air media and use of bacteria and algae). Recently, due to increasing interest in zero-waste cycles that allow for minimal environmental footprints, these technologies have come back into focus as alternative, sustainable solutions for growing food on earth.

Due to space travel constraints, the solutions developed were based on mimicking natural food webs (Fig. 1.8), waste and energy cycles, often using microalgae at the base of the cycle.

Bacteria   Zooplankton   Small Fish   Large Fish   Human
Microalgae

Fig. 1.8 Simplified natural food-web

Figure 1.9 shows an example of an early closed-loop system concept for space travel. This solution was described by British landscape architect Ian McHarg [16], who became known for his nature-inspired urban designs and for assembling some of the first interdisciplinary urban design teams. McHarg's simplified depiction of the system draws parallels to natural cycles, using algal cultures as base elements for air, water, food, and waste cycles.

Regarding aquatic food production, closed-loop food systems have been widely explored in aquaponics, which is focused on freshwater cultures involving multiple species. Recent developments and emergent fields start looking at alagaeponics, maraponics, and haloponics, which are largely based on saline water cultures using aquaponic principles (Fig. 1.10).

Aquaponics is a near closed-loop system, where wastewater from fish species grown in a recirculating aquaculture system (RAS) is reused as fertilizer for hydroponic crops (plants grown in a water medium). By extracting nutrients from fish wastewater, the hydroponic system effectively acts as a filter, and the resulting water can be reused as clean water for fish rearing. The main drawbacks of these systems are the requirement of additional fish feed and nutrient adjustment for plant and fish species growth efficiency.

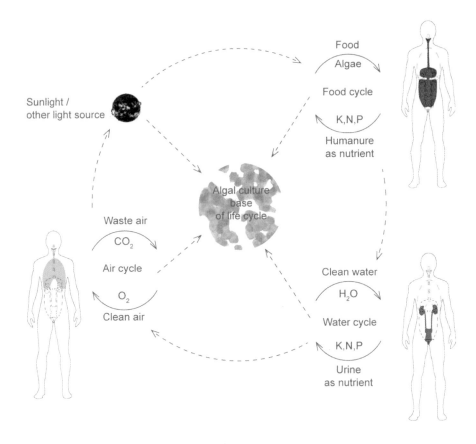

Fig. 1.9 McHarg's closed-loop system for human space transport (based on [16])

Optimizing the nutrient flows is therefore a key parameter in achieving optimal results that can be used for industrial production. To address the issue of different required nutrient states and uptake rates between terrestrial plants grown in the hydroponic system and aquatic species, recent solutions focus on so-called decoupled, multiple loop systems. In these systems, RAS and hydroponic loops are separated with an additional mineralization loop introduced. The mineralization loop allows for nutrient optimization through processes such as desalination [17].

Aside from the issue of nutrient uptake rates, there has been a growing focus on integrating additional species into aquaponic loops to eliminate the need for additional fish feed sources. Here, new applications for micro- and macro-algae cultures from the fields of energy production, desalination, and water treatment sectors have provided potentially viable ways of closing the nutrient loop.

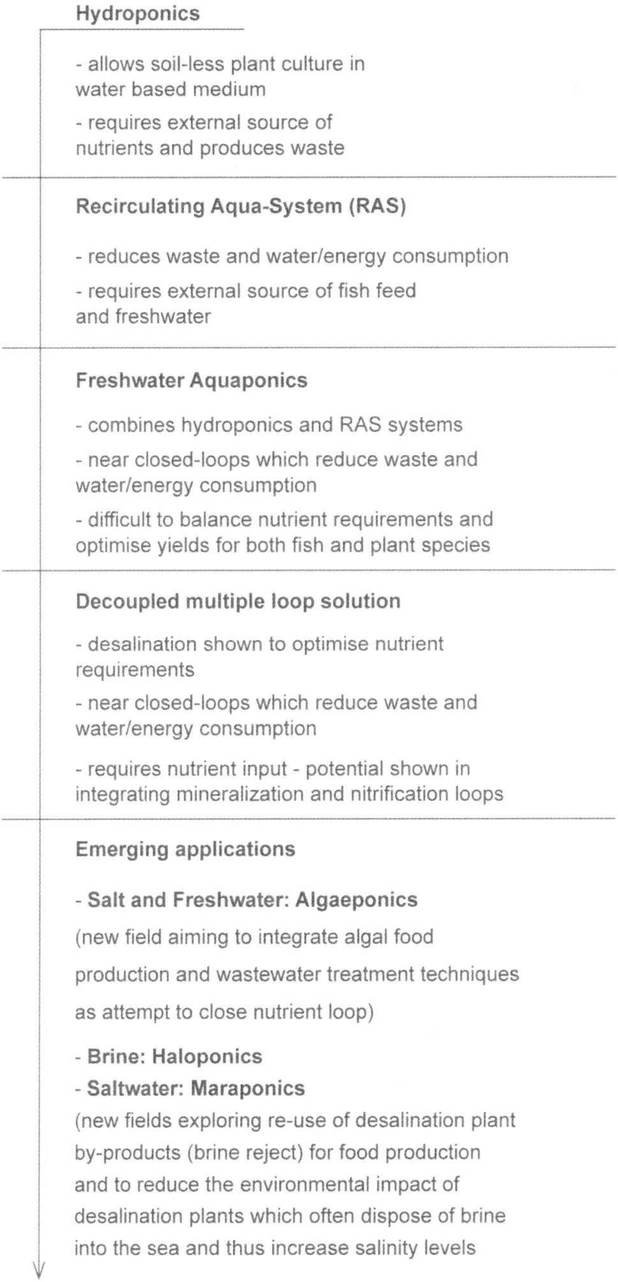

Hydroponics

- allows soil-less plant culture in
water based medium
- requires external source of
nutrients and produces waste

Recirculating Aqua-System (RAS)

- reduces waste and water/energy consumption
- requires external source of fish feed
and freshwater

Freshwater Aquaponics

- combines hydroponics and RAS systems
- near closed-loops which reduce waste and
water/energy consumption
- difficult to balance nutrient requirements and
optimise yields for both fish and plant species

Decoupled multiple loop solution

- desalination shown to optimise nutrient
requirements
- near closed-loops which reduce waste and
water/energy consumption
- requires nutrient input - potential shown in
integrating mineralization and nitrification loops

Emerging applications

- **Salt and Freshwater: Algaeponics**

(new field aiming to integrate algal food

production and wastewater treatment techniques

as attempt to close nutrient loop)

- **Brine: Haloponics**
- **Saltwater: Maraponics**
(new fields exploring re-use of desalination plant
by-products (brine reject) for food production
and to reduce the environmental impact of
desalination plants which often dispose of brine
into the sea and thus increase salinity levels

Fig. 1.10 Development and optimization of near closed-loop aquatic food systems [18]

Microalgae and bacteria consortiums have been efficiently used to treat and retrieve nutrients from municipal wastewater, with the added benefit of harvesting the algae for biofuel production [19]. Conversely, some micro and macro-algae species have been established as viable food sources for fish feed and even human consumption [20,21]. Additionally, in conjunction with marine microalgae, brine reject, and saltwater have been successfully used to irrigate salt-tolerant (halophyte) plant species [22].

In conclusion, although a solution for fully closed-loop production systems is not yet apparent, the combination of algal cultures with decoupled aquaponics and integration of uses from other industries achieves near-closed loop cycles with very little external input, small environmental footprints and added economic values due to product diversification.

In the context of achieving global food security by targeting traditional farmers (smallholders), hybrid solutions, which merge the high efficiency of industrial applications and the simplicity of traditional closed-loop production systems, could provide valuable opportunities for the sustainable development of future food production systems.

1.5 From Problem to Solution

As summary of the first chapter, the following problem and hypotheses have been identified:

The Problem: There is a lack of global food supply because the world population is growing, whereas agricultural land areas cannot be sustainably expanded (Sect. 1.1).

Hypothesis A: Shifting the focus from the land to the water surface, while targeting regions most affected by food insecurity, can create additional food production areas (Sect. 1.2).

Hypothesis B: To make the supply of nutrition accessible, stable, diverse, and effective, solutions for aquatic food production should target smallholders as key drivers of change in food insecure areas (Sect. 1.3).

Hypothesis C: Hybrid closed-loop systems which focus on the transfer of industrial technology into traditional food production systems will make aquatic food production more sustainable by increasing efficiency, reducing fertilizer use and environmental pollution (Sect. 1.4).

Hypotheses A-C, the main outputs of chapter 1, represent the key research directions identified through problem analysis, data collection, and evaluation of state of the art food production technologies.

The next chapters showcase the exploration and proof of novel ideas through the development of an aquatic food system design proposal. How can the hypotheses be utilized to find a solution for the problem? And how can the solution be tested afterwards to prove its significance? To answer these questions, a design and testing method, which may create solutions by transferring traditional terrestrial food- producing systems onto the water, has been developed (see Fig. 1.11).

Subsequently, chapter 2 will analyze a successful model of a traditional closed-loop system driven by smallholders, before shifting the model from land to sea. The result is a design solution which explores how traditional farming principles, closed loop industrial food systems and low-cost manufacturing can be combined to achieve an efficient sea-based farming solution centered on smallholders.

The final chapter 3 will focus on potential applications and procurement routes, providing an assessment of the economic feasibility of the proposed solutions. This will allow for an appraisal regarding the initial question "How far are the created solutions able to counterbalance an increased effort of aqua-cultural food production with its higher efficiencies?"

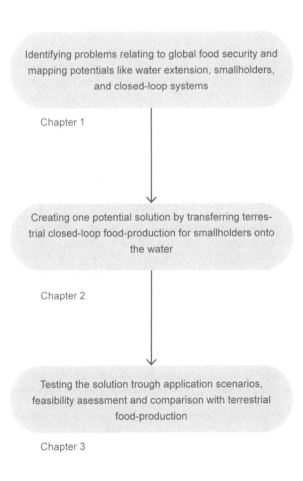

Fig. 1.11 Step-by-step research design logic

References

1. FAO, IFAD, UNICEF, WFP and WHO (2020) The State of Food Security and Nutrition in the World 2020.Transforming food systems for affordable healthy diets. https://doi.org/10.4060/ca9692en
2. United Nations, Department of Economic and Social Affairs, Population Division (2019) World population prospects 2019: highlights (ST/ ESA/SER.A/423)
3. Demand calculated as population $\times$ average per capita demand in ha (49.5% 10.4 bln. Ha/7 bln. People in 201L 0.74 ha/person) with percent land based on: Our World in Data (2011) Share of global habitable land needed for agriculture. https://ourworldindata.org/grapher/share-of-global-habitable-land-needed-for-agriculture-if-everyone-had-the-diet-of?tab=chart&time=latest. Accessed 03 Jan 2022
4. Our World In Data (2019) Global land use for food production. https://ourworldindata.org/land-use#breakdown-of-global-land-use-today. Accessed 03 Jan 2022
5. Koyande A, Chew K, Rambabu K, Tao Y, Chu D, Show P (2019) Microalgae: a potential alternative to health supplementation for humans. Food Sci Human Wellness 8(1):16–24. https://doi.org/10.1016/j.fshw.2019.03.001
6. FAO (2020) The State of World Fisheries and Aquaculture 2020. Sustainability in action. https://doi.org/10.4060/ca9229en
7. NASA (2015) Seaweed farms in South Korea. https://eoimages.gsfc.nasa.gov/images/imagerecords/85000/85747/korea_oli_2014031_lrg.jpg. Accessed 10 Jan 2022
8. FAO (2008) Food security information for action practical guidelines—an introduction to the basic concepts of food security. http://www.fao.org/3/al936e/al936e00.pdf. Accessed 05 Jan 2022
9. UN General Assembly (2015) Transforming our world: the 2030 agenda for sustainable development, A/RES/70/1. http://www.un.org/ga/search/view_doc.asp?symbol=A/RES/70/1&Lang=E. Accessed 05 Jan 2022
10. Based on Grebmer K, Bernstein J, Alders R et al (2020) 2020 Global hunger index: one decade to zero hunger: linking health and sustainable food systems. https://www.globalhungerindex.org/pdf/en/2020.pdf. Accessed 05 Jan 2022
11. FAO and IFAD (2019) United Nations decade of family farming 2019–2028. Global Action Plan. https://www.fao.org/3/ca4672en/ca4672en.pdf. Accessed 05 Jan 2022
12. Ricciardi V, Ramankutty N, Mehrabi Z, Jarvis L, Chookolingo B (2018) How much of the world's food do smallholders produce? Glob Food Sec 17:64–72. https://doi.org/10.1016/j.gfs.2018.05.002
13. FAO Committee on World Food Security (2019) Connecting smallholders to markets. https://www.fao.org/cfs/workingspace/workstreams/past-workstreams/smallholders/en/. Accessed 05 Jan 2022
14. European Commission Joint Research Center, Smallholder Agriculture—Global distribution of farm field size map. https://wad.jrc.ec.europa.eu/smallholderagriculture. Accessed 05 Jan 2022
15. FAO (2013) Fisheries and aquaculture technical paper 580: social and economic dimensions of carrageenan seaweed farming. https://www.fao.org/3/I3344E/i3344e.pdf. Accessed 05 Jan 2022
16. Based on system description provided in Damery D, Webb J, Danylchuk A, Hoque S (2012) Natural systems in building integrated aquaculture design. WIT Trans Ecol Environ 160:87–93. https://doi.org/10.2495/DN120081
17. Goddek S, Keesman KJ (2018) The necessity of desalination technology for designing and sizing multi-loop aquaponics systems. Desalination 428:76–85. https://doi.org/10.1016/j.desal.2017.11.024
18. Goddek S, Joyce A, Kotzen B, Burnell G (eds) (2019) Aquaponics food production systems: combined aquaculture and hydroponic production technologies for the future, 1st edn. Springer International Publishing. https://doi.org/10.1007/978-3-030-15943-6

19. Jia H, Yuan Q (2016) Removal of nitrogen from wastewater using microalgae and microalgae–bacteria consortia. Cogent Environ Sci 2(1):1275089. https://doi.org/10.1080/23311843.2016.1275089

20. Garcia-Vaquero M, Hayes M (2016) Red and green macroalgae for fish and animal feed and human functional food development. Food Rev Intl 32(1):15–45. https://doi.org/10.1080/87559129.2015.1041184

21. Taelman SE, De Meester S, Roef L, Michiels M, Dewulf J (2013) The environmental sustainability of microalgae as feed for aquaculture: a life cycle perspective. Biores Technol 150:513–522. https://doi.org/10.1016/j.biortech.2013.08.044

22. Sánchez AS, Nogueira IBR, Kalid RA (2015) Uses of the reject brine from inland desalination for fish farming, spirulina cultivation, and irrigation of forage shrub and crops. Desalination 364:96–107. https://doi.org/10.1016/j.desal.2015.01.034

Chapter 2
Design

2.1 The Oasis system

For thousands of years, ancient man-made desert oases demonstrate the successful application of closed-loop farming systems [1,2]. They are traditionally driven by small "individual holdings, often less than 1ha" [3] and are a model for resilience and adaptation in the context of current climatic changes and their associated impacts on crops (soil salinization, sea level rise and desertification). Oases have evolved into hybrid farming systems which blend human intervention with the intrinsic qualities of site-adapted plant species to achieve highly efficient resource use within harsh environmental conditions.

The oasis is a closed-loop system centered on date palm cultures. The system is enabled by artificial water harnessing infrastructure and zero waste cycles that utilize human and animal waste as fertilizer. Palm trees can survive in highly saline soil and are adapted to the high solar radiation and temperature differences specific to the desert climate. They are a source of food and building material and enable a degree of control over the local micro-climate.

Shading from palm trees filters solar radiation and cools, to create a suitable habitat for humans and other plant species used for human and livestock consumption. In conjunction with the artificial water harnessing infrastructure, through irrigation, channel bank stabilization and local cooling, palms play an essential role in efficient water resource use by decreasing evaporation and allowing for water collection via condensation and percolation.

Additionally, oases act as connecting network nodes along trade and travel routes. Historically, this allowed oasis settlements to flourish and grow into thriving economic hubs (e.g. along incense and spice routes).

In summary, the four main underlying principles identified (Fig. 2.1) are: closed- loop zero-waste system; use of human water management to enable the growth of adapted species; locally controlled micro-climate for efficient resource use and crop diversification; the creation of networks linked by trade and travel routes.

© The Author(s), under exclusive license to Springer Nature Singapore Pte Ltd. 2023
J. Baumeister and I. C. Giurgiu, *SeaOasis*, SpringerBriefs in Architectural Design
and Technology, https://doi.org/10.1007/978-981-19-1373-0_2

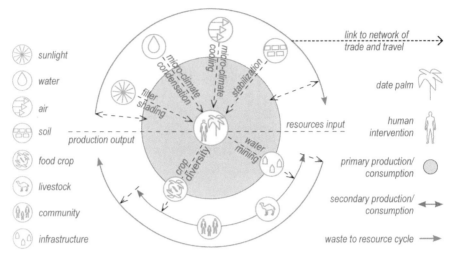

Fig. 2.1 Oasis system principles [1, 2]

2.2 From the Oasis to SeaOasis

Shifting - or transferring - the oasis system principles onto the water results in a conceptual model which we call "SeaOasis". It is a water-based farm that explores how the underlying principles of desert oases can be synthesized into a coherent closed-loop sea-farming system.

The farm is destined for smallholders and can include cultures of high-value micro-algae and halophytes. By providing a way for small farmers to produce these high-value products, as well as by-products that can be consumed locally or sold for additional revenue. SeaOasis aims to boost the economic power of family farms and implicitly increase food security (see Sect. 3.4).

Switching from land-based to marine agriculture systems has the benefit of providing additional crop areas for coastal locations with limited agricultural land availability and may increase the productivity and efficiency of the farm via the high productivity of marine biomass producers. In terrestrial ecosystems, "evolution has favored woody and stem structures in plants, to help them rise above their competitors for light in the absence of water's buoyancy. These structures are heavy, relatively inaccessible to consumers, and make up the bulk of terrestrial plant biomass".[4] Additionally, the growth of biomass on land is the result of long-term processes. By comparison, biomass producers in the ocean have evolved to adapt to the ocean's highly mobile nutrient flows, thus selecting for small sizes and high productivity rates with a fast turnover.[4]

Therefore, the role of key species adapted to the marine environment is somewhat different from that of the date palm. The date palm functions of providing structural stability and local cooling for micro-climatic control are replaced by the buoyancy

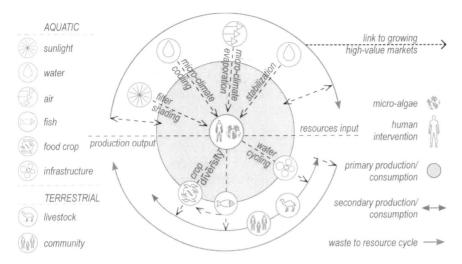

Fig. 2.2 SeaOasis system principles

and cooling effects of seawater. The role of supporting crop diversity, however, remains. In the case of microalgae, this is achieved via food webs and cycles rather than through habitat provision and shading, which in this case are supplied via human intervention. Additionally, like in the desert oases, waste flows from humans and livestock can be re-used as fertilizers, with the farm system's added benefit of having a potentially positive impact on both aquatic and terrestrial systems (Fig. 2.2).

Lastly, while the oases of the desert functioned as nodes along trade and travel routes, in the modern context of marine farming, SeaOases can act as local and global economic and social connectors.

The culture of microalgae for a potent global market with applications in food, pharmaceutical, biofuel, and other industries can aid to establish connections between local farmers and global trade networks.

At a smaller scale, products such as fish and halophytes can establish connections with local markets. At the same time, in some traditional oases, small individual holdings (cultivated areas) can be described as interconnected network nodes (Fig. 2.3). This characteristic is reflected via a modular approach in the SeaOasis concept (Fig. 2.4)

In summary, the underlying principles of the land-based oasis have been adapted to the marine environment in the following ways:

- Closed-loop zero-waste system achieved via enclosed infrastructure and the use of closed-loop nutrient cycles by forming cultured food webs which include key high-value species.

- Use of human water management to facilitate the growth of key adapted species by cycling water to enable the growth of marine micro-algae, fish, and halophytes as key species.

- Controlled local micro-climate for efficient resource use and crop diversification by using seawater and wind for cooling and evaporation, diversifying crops by growing food webs based on micro-algae, and by efficient use of resources via human intervention (design and management).

- Creation of networks linked by trade and travel routes via high-value products for global markets and by-products for local markets.

Aiming to improve current food production systems and achieve sustainable future growth, the application of these principles was tested within a design scenario which combines the SeaOasis concept with the constraints and opportunities of the key research directions identified in the initial analysis of global food systems (see Chap. 1).

The following sections describe the design output of the research, providing an example of how the SeaOasis principles can be applied within a practical design solution that can be implemented by smallholders (with limited digital and technical literacy), yields high values species, and combines high-tech industrial and traditional farming techniques in an innovative way.

Fig. 2.3 Adjir Oasis, Adrar, Algeria: date palm trees grow in a network of artificial hollows. (*Photo* George Steinmetz)

Section 2.3 describes the overall closed-loop system and components, combining both traditional and state of the art hydroponic and aquaponic configurations within a design that strives for simplicity and ease of use.

Microalgae cultures were identified as high-value species and very good candidates for several coupled systems aiding to achieve efficient closed-loop systems (see Sect. 1.4). Section 2.4 presents several possible species groups that can be cultured as closed cycles, each yielding high-value algae species.

Additionally, to achieve efficiencies of algal cultures that can be used in industrial applications, a review of existing industrial photobiorector designs was conducted with the aim of identifying a suitable reactor that can be adapted to fit the design criteria for submerged applications, easy maintenance, and use. Section 2.5 succinctly presents the findings of this review: background of existing applications; state of the art photobioreactor design; and selection of an adaptable reactor to integrate within the SeaOasis components.

The integration of the selected photobioreactor as well as further proposals for how the design can be enhanced in order to optimize solar exposure and enhance algal growth are presented in section 2.6.

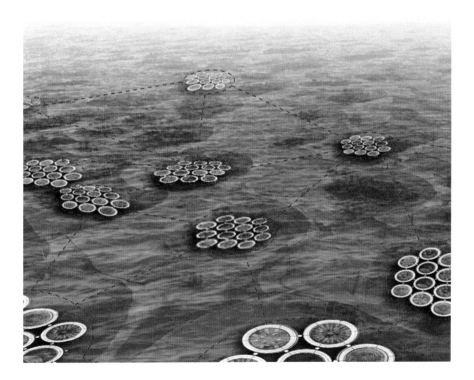

Fig. 2.4 Concept of a SeaOasis farm network

2.3 The Aqua-Pods

The SeaOasis consists of aqua-pods that allow the sustainable culture of fish, algae, and plants in either fresh, brackish, or seawater mediums. It integrates several existing systems and emerging strands of research into a marine farming unit. As the SeaOasis farm is designed for the marine environment, the most straightforward configuration utilizes the readily available seawater as a resource and consists of five primary floating sub-units that together form water, nutrient, and waste recycling loops as a (nearly) closed-loop system.

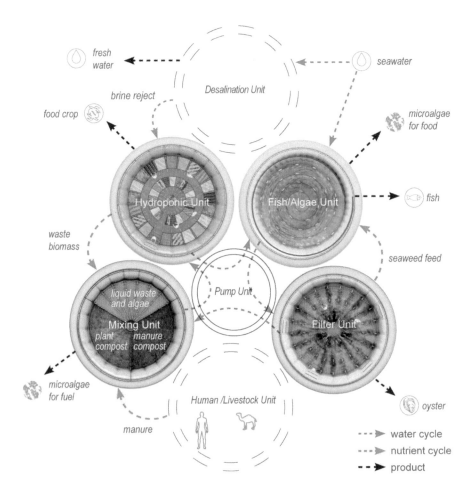

Fig. 2.5 Aqua-pods enabling SeaOasis processes

The **Fish/Algae Unit** (Fig. 2.5 clockwise from top right) combines submerged, closed algae photobioreactors with a fish growing tank. Fish wastewater from this tank is rich in nutrients and can grow macro-algae, with sufficient nutrient removal that the resulting water can be released back into the sea [5]. Additionally, bivalves (e.g. oysters and clams) are known to filter solid fish waste efficiently [6].

The **Filter Unit** uses macro-algae and bivalve cultures which efficiently remove both liquid and solid fish wastes from the water. Cultured seaweed can be used as feed for the fish tanks, while bivalves represent high-value products that can be monetized.

Clean water from the **Filter Unit** is subsequently fed into the **Mixing Unit** that combines the resulting sludge from composting solid plant waste and manure with algae-treated human liquid wastewater to obtain a liquid fertilizer solution. Algae cultures from this unit can be harvested and monetized as biomass for biofuel production.

The resulting liquid fertilizer is further diluted with seawater and fed into **the Hydroponic Unit,** consisting of several deep water culture beds. For the seawater scenario, salt-adapted plants such as Salicornia, which can be further used for human consumption, biochar, or livestock feed can be grown [7]. Further to nutrient uptake by the plants in this unit, water can be recycled into the fish-growing tanks with limited supplemental seawater needed.

Lastly, the **Pump Unit** is a technical space unit that houses a small generator and moves water from one tank to the next. Given the farm's location, unobstructed access to solar energy provides a possible solution to generate power for the pumps without using fossil fuels.

Depending on the algae, fish, and plant species selected, the system can be adapted to accommodate freshwater cultures via on or off-site desalination.

The Hydroponic Unit (Fig 2.6) consists of a base unit and rope and buoy system, which provide the structure for attaching locally constructed floating plant beds similar to those seen in traditional floating garden (Dhap) systems in Bangladesh [8]. Depending on the overall system configuration, the unit can be used for fresh or saltwater cultures, which are used for local consumption.

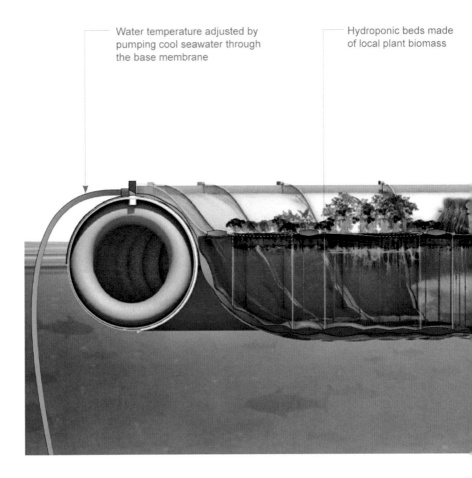

Water temperature adjusted by pumping cool seawater through the base membrane

Hydroponic beds made of local plant biomass

Fig. 2.6 The hydroponic unit

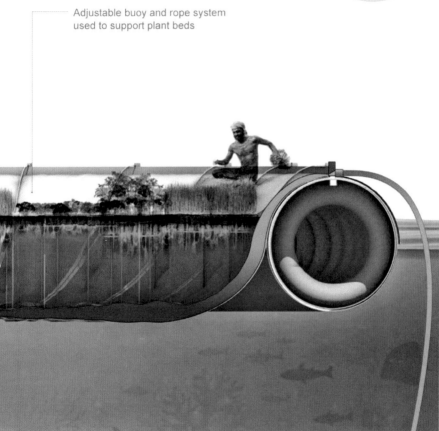

Adjustable buoy and rope system
used to support plant beds

The Filter Unit (Fig. 2.7) consists of a base unit and rope and buoy system, which provide the vertical structure for suspended oyster cultures and horizontal structure for macro-algae cultures. In open oyster and mussel farms, seaweed and bivalve polycultures have been shown to have increased yields, higher market returns and lower environmental footprints due to the intrinsic natural fertilization processes [9].

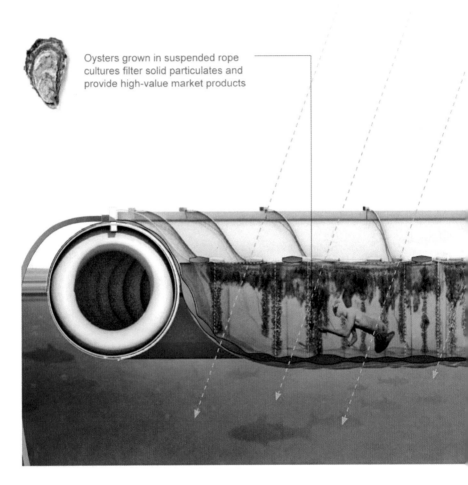

Oysters grown in suspended rope cultures filter solid particulates and provide high-value market products

Fig. 2.7 The filter unit

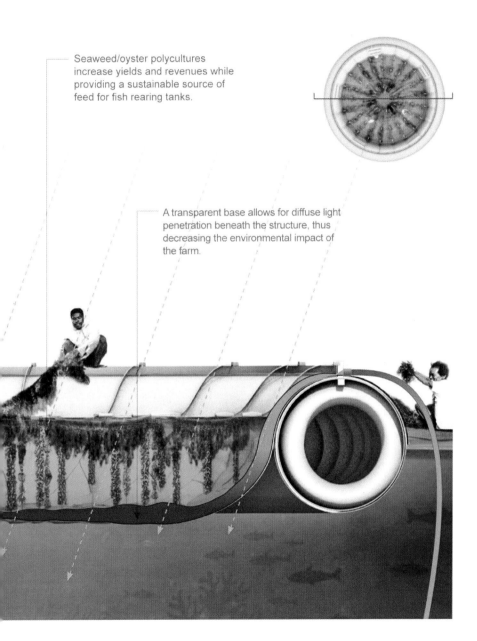

Seaweed/oyster polycultures increase yields and revenues while providing a sustainable source of feed for fish rearing tanks.

A transparent base allows for diffuse light penetration beneath the structure, thus decreasing the environmental impact of the farm.

The Fish / Algae Unit (Fig. 2.8) comprises all the modular components (base unit, rope and buoy and algae reactor systems). As the algae reactors are enclosed, the species grown can be selected and mixed to maximize profits. Furthermore, the modular reactor system allows for bags to be individually and easily removed for harvest or replaced. This allows operational risk mitigation and a reduction in overall maintenance and labour costs.

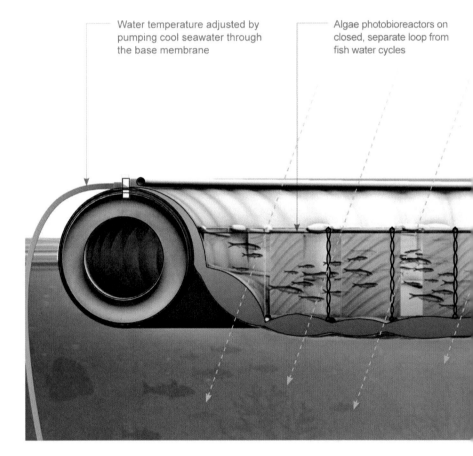

Water temperature adjusted by pumping cool seawater through the base membrane

Algae photobioreactors on closed, separate loop from fish water cycles

Fig. 2.8 The fish/algae unit

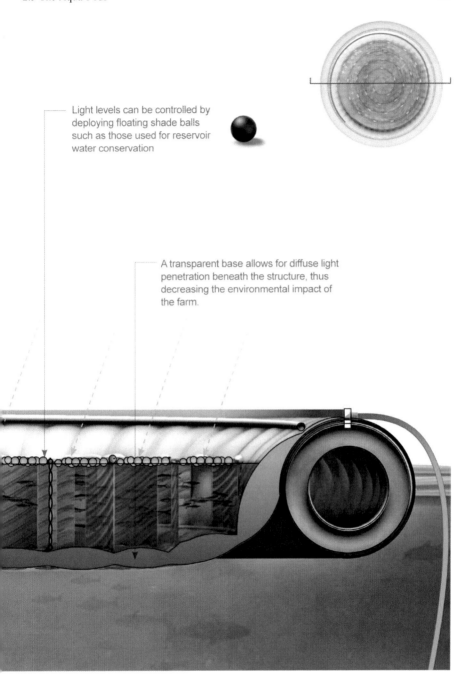

Light levels can be controlled by deploying floating shade balls such as those used for reservoir water conservation

A transparent base allows for diffuse light penetration beneath the structure, thus decreasing the environmental impact of the farm.

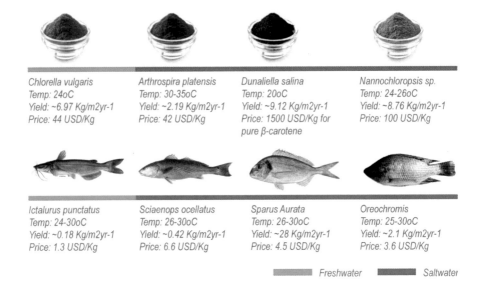

Chlorella vulgaris	Arthrospira platensis	Dunaliella salina	Nannochloropsis sp.
Temp: 24oC	Temp: 30-35oC	Temp: 20oC	Temp: 24-26oC
Yield: ~6.97 Kg/m2yr-1	Yield: ~2.19 Kg/m2yr-1	Yield: ~9.12 Kg/m2yr-1	Yield: ~8.76 Kg/m2yr-1
Price: 44 USD/Kg	Price: 42 USD/Kg	Price: 1500 USD/Kg for pure β-carotene	Price: 100 USD/Kg

Ictalurus punctatus	Sciaenops ocellatus	Sparus Aurata	Oreochromis
Temp: 24-30oC	Temp: 26-30oC	Temp: 26-30oC	Temp: 25-30oC
Yield: ~0.18 Kg/m2yr-1	Yield: ~0.42 Kg/m2yr-1	Yield: ~28 Kg/m2yr-1	Yield: ~2.1 Kg/m2yr-1
Price: 1.3 USD/Kg	Price: 6.6 USD/Kg	Price: 4.5 USD/Kg	Price: 3.6 USD/Kg

▬▬▬ Freshwater   ▬▬▬ Saltwater

2.4 Potential Species

As the proposed aqua-pod units work together to form closed-loop cycles, each coupled system's sub-species must be appropriately selected. Optimal growth temperature and salinity ranges for fish/algae and oyster/seaweed couples must be matched between species. To reduce energy consumption, they must also match the typical site conditions (e.g. optimal temperature ranges should overlap with average sea surface temperature).

The examples shown below (Fig. 2.9/2.10) indicate a few of the more commonly cultured species that could be coupled in freshwater and salt-water cultures. To estimate the proposal's economic feasibility, yields/m2 and market selling prices are shown for algae and oyster species which are the two main high-value products. Seaweed, fish, and legumes can also be considered for local markets, but we have assumed that they are either used for farmers' consumption, as livestock or as fish feeds within SeaOasis processes.

2.5 Photobioreactors

Although the aqua-pod system yields various products, micro-algae cultures represent a central element of the artificial food web due to the high value of market selling prices and pivotal role in nutrient removal cycles. In order to match the efficiency of industrial applications, it is proposed that the aqua-pod units include an algae growth system adapted from industrial applications.

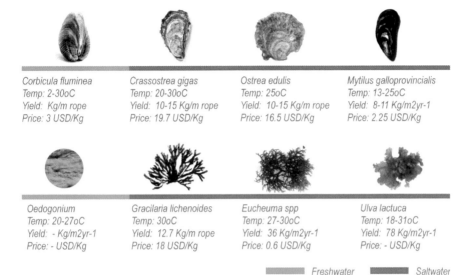

Corbicula fluminea
Temp: 2-30oC
Yield: Kg/m rope
Price: 3 USD/Kg

Crassostrea gigas
Temp: 20-30oC
Yield: 10-15 Kg/m rope
Price: 19.7 USD/Kg

Ostrea edulis
Temp: 25oC
Yield: 10-15 Kg/m rope
Price: 16.5 USD/Kg

Mytilus galloprovincialis
Temp: 13-25oC
Yield: 8-11 Kg/m2yr-1
Price: 2.25 USD/Kg

Oedogonium
Temp: 20-27oC
Yield: - Kg/m2yr-1
Price: - USD/Kg

Gracilaria lichenoides
Temp: 30oC
Yield: 12.7 Kg/m rope
Price: 18 USD/Kg

Eucheuma spp
Temp: 27-30oC
Yield: 36 Kg/m2yr-1
Price: 0.6 USD/Kg

Ulva lactuca
Temp: 18-31oC
Yield: 78 Kg/m2yr-1
Price: - USD/Kg

Freshwater   Saltwater

The industrial applications of micro-algae range from food products to wastewater treatment and biofuels, and significant advances have been made in the last decade to optimize algae culture beds for optimal light availability and yield.

In wastewater treatment applications, algae are often grown in open raceway ponds. While these systems are efficient in nutrient removal, they require large land areas and are prone to be affected by pathogens and climatic conditions. However, in the food and chemical industry, much more refined control of the growth medium and conditions are required. This has prompted the development of so-called closed photobioreactors (PBR), which are enclosed glass or plastic tubes, or panels used toculture algae in a controlled medium.

Fig. 2.11 Outdoor pilot Proviron microalgae reactors (*Photo* Proviron)

A newer field of research is the area of floating PBRs. Although there are still relatively few studies (compared to land-based applications), preliminary results suggest that the use of closed floating PBR's in aquatic environments presents several benefits that could help save energy and provide low-cost solutions for culturing algae. The natural buoyancy of water supports floating or submerged PBRs automatically, whereas land-based flexible bag solutions require significant infrastructure, making floating PBRs even more efficient.

Because, with depth, algae growth can be inhibited by the uppermost layers of biomass, the cultures require mixing to ensure the best yields. In terrestrial applications, this is either done mechanically or by pumping air bubbles. In the marine environment, the waves' natural motion provides the opportunity for "passive" mixing using wave energy. Thus far, this type of mixing has been tested with good results in horizontal floating reactors [10].

One example of a promising prototype for submerged PBR's is the ProviAPT culture panel (Fig. 2.11 and 2.12), which consists of an "array of vertical flat-panel type reactors enclosed in a translucent plastic bag filled with water"[11]. The panels are designed to maximize light exposure while minimizing cost and were initially tested in the Wageningen University Algae Parc facilities [12], with larger applications being tested for re-use of flue-gas for algal culture in Antwerp (Belgium) [11].

Fig. 2.12 Microalgae vertical farming at Proviron (*Photo* Proviron)

The estimated cost for the ProviAPT rector amounts to a total "not exceeding € 10/m²" [11], which would further be reduced by installing the panels in aquatic environments rather than on land. As the panels are simple to install and require minimal maintenance, this solution could be a viable application for smallholders.

Additionally, due to its' low cost, modularity and potential for integration into the SeaOasis design, the ProviAPT panel reactor design was deemed as an appropriate rector type to be used as the main algae culture reactor of the SeaOasis aqua-pods (see Fig. 2.8).

Furthermore, given appropriate site selection and maximization of solar exposure, the transfer of this type of technology within systems such as SeaOasis could increase the overall productivity of the pods, providing better yields and implicitly higher revenues for farmers in developing regions of the world.

2.6 Optimizing Sun Exposure

The optimal use of natural resources such as sunlight is a key factor for increasing productivity and decreasing maintenance and labour costs. To maximize solar exposure, the attachment system between the aqua-pod modules and submerged PBR (e.g. ProviAPT) can been adjusted to allow a seasonal rotation of the panels (Fig. 2.13).

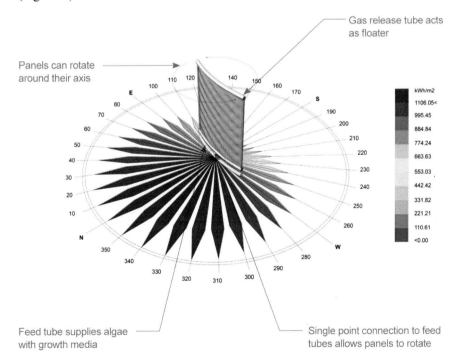

Fig. 2.13 Year-round total available radiation

The simulations (Fig. 2.14/2.15) show a generic scenario for a sub-tropical site in the Southern Hemisphere (latitude 27.94°). Total available radiation and on-panel radiation for the optimal rotation angle have been simulated based on historical climate data. The panels' rotation allows for very similar on-panel radiation levels to be achieved throughout the year.

As plants and algae mostly utilize the visible light spectrum (photosynthetically active radiation) [13], the total radiation available for photosynthesis will be roughly 43% of the total radiation values [14] shown below. Maximizing exposure is, therefore, a key aspect of increasing algae growth and productivity.

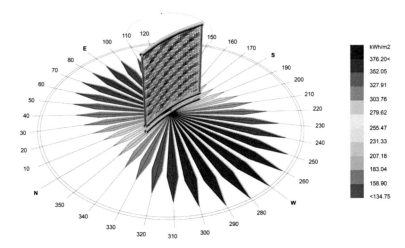

Fig. 2.14 Example: Total radiation on panel (kWh/m2) from Jan 1, 6:00 to Feb 31, 18:00 at latitude 27.94° S

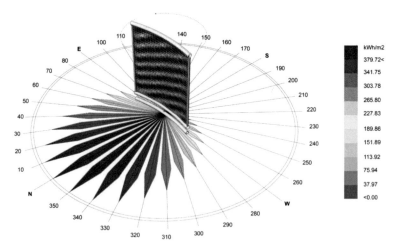

Fig. 2.15 Example: Total radiation on panel (kWh/m2) from Jun 1, 6:00 to Aug 31, 18:00 at latitude 27.94° S

References

1. Baumeister J, Ottmann DA (2015) Urban ecolution—a pocket generator to explore future solutions for healthy and ecologically integrated cities. UWAP, Crawley
2. LabOasis Foundation (2020) Why the Oasis? http://www.laboasis.org/home/why-the-oasis/. Accessed 05 Jan 2022
3. Stevens JH (1972) Oasis agriculture in the Central and Eastern Arabian Peninsula. Geography 57(4):321–326. http://www.jstor.org/stable/40567915
4. Burgess MG, Gaines SD (2018) The scale of life and its lessons for humanity. Proc Natl Acad Sci USA 115(25):6328–6330. https://doi.org/10.1073/pnas.1807019115
5. Neori A, Shpigel M, Ben-Ezra D (2000) A sustainable integrated system for culture of fish, seaweed and abalone. Aquaculture 186(3–4):279–291. https://doi.org/10.1016/S0044-848 6(99)00378-6
6. Strand Ø, Jansen HM, Jiang Z, Robinson SMC (2019) Perspectives on bivalves providing regulating services in integrated multi-trophic aquaculture. In: Smaal A, Ferreira J, Grant J, Petersen J, Strand Ø (eds) Goods and services of marine bivalves. Springer, Cham. https://doi.org/10.1007/978-3-319-96776-9_11
7. Sánchez AS, Nogueira IBR, Kalid RA (2015) Uses of the reject brine from inland desalination for fish farming, spirulina cultivation, and irrigation of forage shrub and crops. Desalination 364:96–107. https://doi.org/10.1016/j.desal.2015.01.034
8. Ministry of Agriculture, People's Republic of Bangladesh (2017) Floating garden agricultural practices in Bangladesh, a proposal for globally important agricultural heritage systems (GIAHS). http://www.fao.org/publications/card/en/c/4c57fc69-e538-46e0-969c-d19 7be845a5f/. Accesses 05 Jan 2022
9. UNDP, FAO Regional Seafarming Project RAS/86/024 (1989) Culture of Kelp (Laminaria japonica) in China—training manual 89/5. http://www.fao.org/3/AB724E/AB724E08.htm. Accessed 05 Jan 2022
10. Zhu C, Han D, Li Y, Zhai X, Chi Z, Zhao Y, Cai H (2019) Cultivation of aquaculture feed Isochrysis zhangjiangensis in low-cost wave driven floating photobioreactor without aeration device. Biores Technol 293:122018. https://doi.org/10.1016/j.biortech.2019.122018
11. Roef L, Jacqmain M, Michiels M (2012) 13 case study: microalgae production in the self-supported ProviAPT vertical flat-panel photobioreactor system. In: Posten C, Walter C (eds) Microalgal biotechnology: potential and production. De Gruyter, Berlin, Boston, pp 243–246. https://doi.org/10.1515/9783110225020.243
12. Hendrik de Vree J (2016) Outdoor production of microalgae. https://edepot.wur.nl/387236. Accessed 05 Jan 2022
13. Fondriest Environmental (2014) What is photosynthetically active radiation? In: Solar radiation and photosynethically active radiation. Fundamentals of environmental measurements. Fondriest Environmental. https://www.fondriest.com/environmental-measurements/par ameters/weather/photosynthetically-active-radiation/#PAR6. Accessed 05 Jan 2022
14. Mõttus M, Sulev M, Baret F, Lopez-Lozano R, Reinart A (2012) Photosynthetically active radiation: measurement and modeling. In: Meyers RA (eds) Encyclopedia of sustainability science and technology. Springer, New York, NY. https://doi.org/10.1007/978-1-4419-0851-3_451

Chapter 3
Application

3.1 Low-Cost Production and Revenue

The SeaOasis' design iteration presented in chapter 2 is further detailed and assessed here via an implementation scenario which follows the principles of affordability, efficiency, and sustainability. Therefore, the design utilizes modular systems with low-cost, readily available materials, aiming to lower production costs by minimizing the number of custom components. At the same time, it allows a simple installation and maintenance that does not require specific training. The system can be assembled on-site by any farmer.

The different aqua-pods consist of three main modular systems (Fig. 3.1), which can be combined in different configurations to obtain each of the main proposed unit types (pump, mixing, filter, and fish/algae). Initial investment cost ranges represent indicative component production prices (see appendix B), excluding pumps and generators, shipping, and starter cultures for each species. For this initial cost estimate, each of the modules (mixing, filter, fish/algae, and hydroponic units) were assumed of similar size (10m diameter rings). Before the implementation, each system should be sized to optimize input and output ratios for each species, depending on the proposed SeaOasis size.

The **Base System** consists of a circular floating rim made from tractor tyre inner tubing and an ETFE or similar base layer, which separates the system from surrounding seawater. The double-layered base system can be heat-welded in various configurations, with seams acting as attachment points for the float system above. This system can be used as a pump unit and has an estimated initial investment cost of 11,500 USD.

The **Float System** acts as a structural grid that allows subsequent attachment of farmed species (e.g. oysters), hydroponic culture beds or algae reactor bags.

J. Baumeister and I. C. Giurgiu, *SeaOasis*, SpringerBriefs in Architectural Design and Technology, https://doi.org/10.1007/978-981-19-1373-0_3

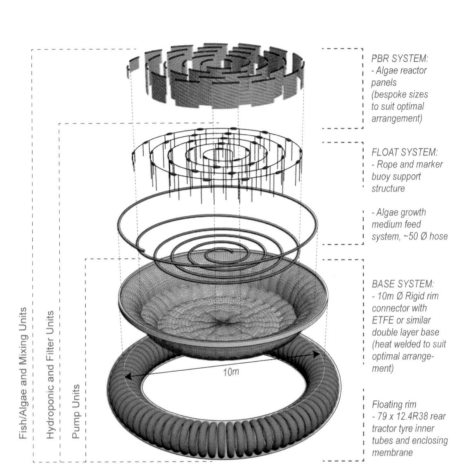

Fig. 3.1 Modular system of the aqua-pods

The system consists of a simple rope and buoy network that attaches to a bottom flexible tube to be installed in different arrangements. Depending on the arrangement, the estimated float system's initial investment cost ranges between 9,000 and 14,000 USD.

The **PBR System** consists of algae culture reactor bags and – used in conjunction with the base and float systems – forms either fish/algae or mixing units, leading to a total initial investment cost of between 21,000 and 26,000 USD per unit (base + float + PBR system cost).

Depending on the chosen float system arrangements, each unit's cost and revenue can be optimized, with each arrangement presenting specific advantages in terms of yields, ease of access, maintenance, and associated risks.

3.2 Flexibility and Optimization

Figures 3.3 and 3.4 show three examples of aqua-pods' arrangements in terms of initial investment cost and potential revenue for each of the high-value species cultured. Different grid structures (Fig. 3.2) offer thereby various opportunities:

The **Circular Grid** is the most compact and allows the largest culture areas of the three scenarios. However, this option implies higher investment costs, more complex connections between feed tubes, and more laborious maintenance and harvesting techniques due to more difficult access to the central rings.

The **Spiral Grid** option is the least expensive in terms of initial investment but is also the least compact and allows fewer culture areas than the other variants. However, this arrangement is the simplest to install, provides excellent access, which reduces labour time and complexity for harvesting and maintenance.

The **Radial Grid**, while the most expensive in terms of initial investment, combines the ease of access and installation of the spiral with the circular option's compact arrangement. Due to the float system's density, this type of arrangement is ideal for hydroponic units that require more buoyancy to support the plant beds.

Circular   Spiral   Radial

Fig. 3.2 Base grid options

Filter Unit

Circular Grid
Oyster line length: ca. 185m
Seaweed line length: ca. 80m
Initial investment: $ 24,000

Spiral Grid
Oyster line length: ca. 130m
Seaweed line length: ca. 65m
Initial investment: $ 20,500

Radial grid
Oyster line length: ca. 160m
Seaweed line length: ca. 70m
Initial investment: $ 25,500

Revenue for oysters
- *Mytilus galloprovincialis*: $ 4,000
- *Corbicula fluminea*: $ 5,500
- *Ostrea edulis*: $ 40,000
- *Crassostrea gigas*: $ 47,500

- *Mytilus galloprovincialis*: $ 3,000
- *Corbicula fluminea*: $ 4,000
- *Ostrea edulis*: $ 27,500
- *Crassostrea gigas*: $ 33,000

- *Mytilus galloprovincialis*: $ 3,500
- *Corbicula fluminea*: $ 5,000
- *Ostrea edulis*: $ 34,500
- *Crassostrea gigas*: $ 41,500

Revenue for seaweed
- *Eucheuma spp.*: $ 2,000
- *Gracilaria lichenoides*: $ 18,000

- *Eucheuma spp.*: $ 1,500
- *Gracilaria lichenoides*: $ 15,000

- *Eucheuma spp.*: $ 1,500
- *Gracilaria lichenoides*: $ 16,000

Fig. 3.3 Filter unit configuration options (rounded revenue in USD/year)

Fish/Algae Unit

Circular Grid
Algae bag area: ca. 60m2
Fish tank area: ca. 50m2;
Initial investment: $ 24,500

Spiral Grid
Algae bag area: ca. 50m2
Initial investment: $ 21,000

Radial grid
Algae bag area: ca. 50m2
initial investment: $ 26,000

Revenue for microalgae
- *Arthrospira platensis*: $ 5,500
- *Chlorella vulgaris*: $ 18,500
- *Nannochloropsis sp.*: $ 52,500
- *Dunaliella salina*: $ 115,000

- *Arthrospira platensis*: $ 4,500
- *Chlorella vulgaris*: $ 14,500
- *Nannochloropsis sp.*: $ 41,000
- *Dunaliella salina*: $ 90,000

- *Arthrospira platensis*: $ 5,000
- *Chlorella vulgaris*: $ 16,000
- *Nannochloropsis sp.*: $ 45,500
- *Dunaliella salina*: $ 99,500

Fig. 3.4 Fish/Algae unit configuration options (rounded revenue in USD/year)

3.3 SeaOasis Configurations

As mentioned in the description of how land-based oases function, each oasis forms a closed-loop, self-sufficient system. Connected by trade and transport routes, land-based oasis form fractal-like, decentralized networks through the arrangement and connection of different loops, which work across various scales to form a larger whole. Similarly, the SeaOasis concept functions based on aqua-pod configurations, which form individual near-closed loops linked through trade via access to growing high-value international markets.

The following configurations (Fig. 3.6/3.7/3.8) combine the base grid arrangement options and associated crop areas described in section 3.2 with examples of selected species couples adapted to the environmental condition bands identified in section 2.4 to highlight potential economic risk levels, revenues and investments associated with each application. Configurations A-E provide a range of mixed salt and freshwater solutions with decreasing investment return times and increasing associated risk levels (Fig. 3.5). Configuration A, therefore, has the lowest risk and highest return time, while configuration E results in the highest risk and lowest return time.

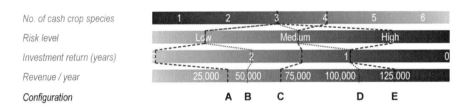

Fig. 3.5 Configurations risk, investment return, and revenue levels of five examples A–E

Generally, higher-value species require more controlled culturing conditions and are often less hardy in terms of disease vulnerability and adaptability to varying conditions, implying more significant risks. Additionally, the number of species used as "cash crops" was considered as a factor that can increase or decrease overall vulnerability and risk levels. Options relying on a single high-value species were deemed higher risk than those that rely on combined incomes from multiple species.

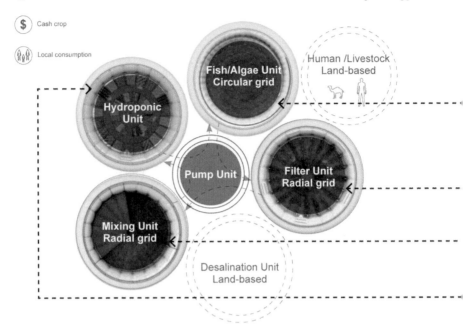

Fig. 3.6 Configuration A: rounded investment, return time, and revenue/year

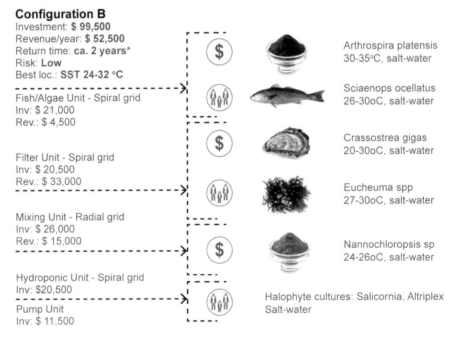

Fig. 3.7 Configuration B and C: rounded investment, return time, and revenue/year

Configuration A
Investment: **$ 113,000**
Revenue/year: **$ 38,500**
Return time: **ca. 3 years***
Risk: **Low**
Best loc.: **SST 24-32 °C**
Proximity to desal. plant

Fish/Algae Unit - Circular grid
Inv: $ 24,500
Rev.: $ 18,500

Filter Unit - Radial grid
Inv: $ 25,500
Rev.: $ 5,000

Mixing Unit - Radial grid
Inv: $ 26,000
Rev.: $ 15,000

Hydroponic Unit - Radial grid
Inv: $25,500

Pump Unit
Inv: $ 11,500

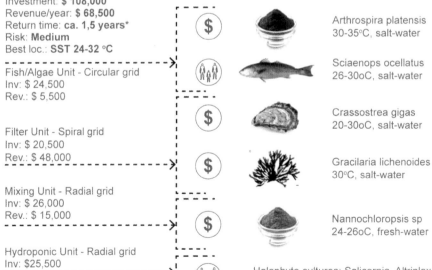

Chlorella vulgaris
24°C, fresh-water

Ictalurus punctatus
24-30°C, fresh-water

Corbicula fluminea
22-30°C, fresh-water

Oedogonium
20-27°C, fresh-water

Nannochloropsis sp
24-26oC, fresh-water

Legumes e.g. tomato basil, lettuce, rice
(Fresh-water)

Configuration C
Investment: **$ 108,000**
Revenue/year: **$ 68,500**
Return time: **ca. 1,5 years***
Risk: **Medium**
Best loc.: **SST 24-32 °C**

Fish/Algae Unit - Circular grid
Inv: $ 24,500
Rev.: $ 5,500

Filter Unit - Spiral grid
Inv: $ 20,500
Rev.: $ 48,000

Mixing Unit - Radial grid
Inv: $ 26,000
Rev.: $ 15,000

Hydroponic Unit - Radial grid
Inv: $25,500

Pump Unit
Inv: $ 11,500

Arthrospira platensis
30-35°C, salt-water

Sciaenops ocellatus
26-30oC, salt-water

Crassostrea gigas
20-30oC, salt-water

Gracilaria lichenoides
30°C, salt-water

Nannochloropsis sp
24-26oC, fresh-water

Halophyte cultures: Salicornia, Altriplex
Salt-water

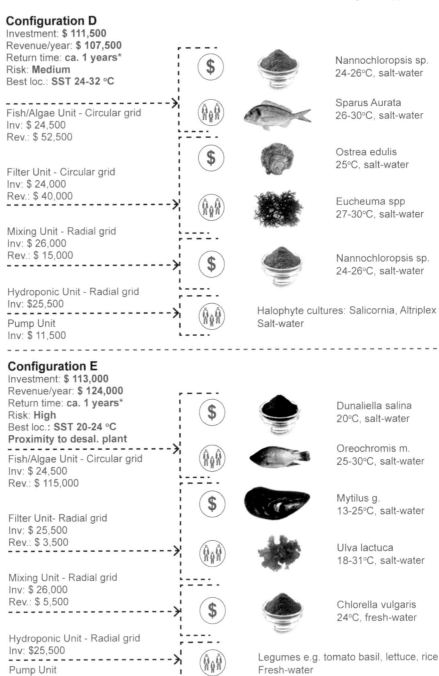

Configuration D
Investment: **$ 111,500**
Revenue/year: **$ 107,500**
Return time: **ca. 1 years***
Risk: **Medium**
Best loc.: **SST 24-32 °C**

Fish/Algae Unit - Circular grid
Inv: $ 24,500
Rev.: $ 52,500

Filter Unit - Circular grid
Inv: $ 24,000
Rev.: $ 40,000

Mixing Unit - Radial grid
Inv: $ 26,000
Rev.: $ 15,000

Hydroponic Unit - Radial grid
Inv: $25,500

Pump Unit
Inv: $ 11,500

Nannochloropsis sp.
24-26°C, salt-water

Sparus Aurata
26-30°C, salt-water

Ostrea edulis
25°C, salt-water

Eucheuma spp
27-30°C, salt-water

Nannochloropsis sp.
24-26°C, salt-water

Halophyte cultures: Salicornia, Altriplex
Salt-water

Configuration E
Investment: **$ 113,000**
Revenue/year: **$ 124,000**
Return time: **ca. 1 years***
Risk: **High**
Best loc.: **SST 20-24 °C**
Proximity to desal. plant

Fish/Algae Unit - Circular grid
Inv: $ 24,500
Rev.: $ 115,000

Filter Unit- Radial grid
Inv: $ 25,500
Rev.: $ 3,500

Mixing Unit - Radial grid
Inv: $ 26,000
Rev.: $ 5,500

Hydroponic Unit - Radial grid
Inv: $25,500

Pump Unit
Inv: $ 11,500

Dunaliella salina
20°C, salt-water

Oreochromis m.
25-30°C, salt-water

Mytilus g.
13-25°C, salt-water

Ulva lactuca
18-31°C, salt-water

Chlorella vulgaris
24°C, fresh-water

Legumes e.g. tomato basil, lettuce, rice
Fresh-water

Fig. 3.8 Configuration D and E: rounded investment, return time, and revenue/year

3.4 SeaOases Communities

The UN World Food Programme has initiated several smallholder market support initiatives [1] which aim to improve market access for smallholder farmers locally and globally. To address local small-scale production and supply and global large- scale demands, it is proposed that the five configurations A-E are implemented within communities of roughly 100 people, with a communal collection of specific high-value products which would be marketed through UN Food Programme initiatives.

Considering an average family size of five people with each family maintaining two SeaOasis configurations of the same type (Aqua-pods for Fish/Algae + Filter + Mixing + Hydroponic + Pump), each community would utilize ca. 40 SeaOases. Due to the pod size of 10m Ø and the suggested configuration of maximum three pods in each direction, the footprint would be around 0.1ha for each SeaOasis family and 2ha for a SeaOases community. Figure 3.9 shows regional national averages for family farm sizes, household sizes, and yearly SeaOasis family incomes.

Each SeaOasis configuration implies varying degrees of associated risks, initial investment return times, and revenues. Given that farmers in areas with high food insecurity are more likely to require lower risk and initial investment costs, it is

	avg. farm size	avg. household	avg. yearly income	proximity*
Latin America and Caribbean	4.8 ha	5.2 people	$ 46,617	
Sub-Sahran Africa	1.8 ha	5.6 people	$ 6,928	
Asia	0.8 ha	5.1 people	$ 3,808	
SeaOasis configuration A	0.1 ha	5 people	$ 77,000	❦ ●
SeaOasis configuration B	0.1 ha	5 people	$ 105,000	❦ ❦
SeaOasis configuration C	0.1 ha	5 people	$ 138,000	❦
SeaOasis configuration D	0.1 ha	5 people	$ 215,000	
SeaOasis configuration E	0.1 ha	5 people	$ 248,000	●

*In proximity to locations with global hunger indexes: [3] ❦ Alarming ❦ Serious ❦ Moderate

and to existing reverse osmosis plants [4] ●

Fig. 3.9 Income of smallholder SeaOasis families in comparison to land-based smallholder agriculture [2]

proposed that configurations A and B are implemented in coastal areas adjacent to countries with alarming or serious global hunger indexes. Configuration C is proposed for areas with moderate global hunger index ratings and configurations D and E for areas with low hunger index ratings. Additionally, as the mixed fresh and saltwater configurations (A and E) utilize brine reject and freshwater, these types of systems have been located in proximity to existing coastal desalination plants (Fig. 3.15).

The comparison of farm sizes and family income between SeaOasis families and land-based smallholders (Fig. 3.9) in Latin America, the Caribbean, sub-Saharan Africa, and Asia answers the initial question regarding the efficiency of sustainable aqua-cultural food production in an imposing way, demonstrating the viability and ample potential of the SeaOasis.

3.5 Strategic Siting

The strategic siting of the SeaOases is an essential factor that ultimately determines the achievable yields for each cultured species and significantly impacts costs and risks associated with labour, maintenance, and overall energy consumption. For example, in the case of wall panel photobioreactors, a 2016 study comparing prices and productivity between two Mediterranean locations (Italy and Tunisia) [5] showed that Tunisia's more favorable conditions would lead to a 50% increase in productivity and 50% decrease in associated costs.

Therefore, strategic siting must support smallholder farmers to locally reduce food insecurity while, at the same time, optimize associated yields, costs, and risks by targeting locations where the environmental context offers the best crop growth conditions.

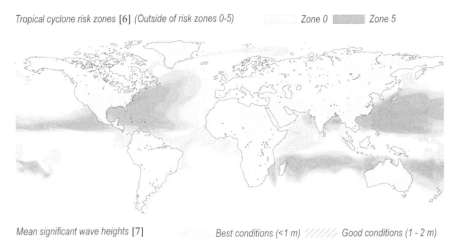

Fig. 3.10 Annual mean significant wave height (m) and tropical cyclone risk zones

Best conditions - constant year round (in °C)   24-32   20-24

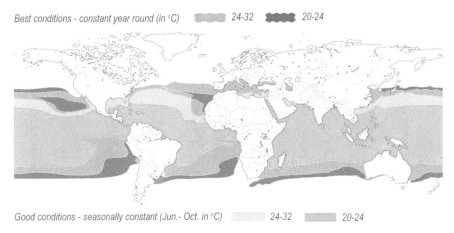

Good conditions - seasonally constant (Jun.- Oct. in °C)   24-32   20-24

Fig. 3.11 Average sea surface temperature [8]

Environmental factors for best and very good growth conditions include wave height and non-occurrence of cyclones, as well as sea surface temperatures and radiation:

Figure 3.10 highlights best and very good conditions for the factors of wave height and storm intensities which represent critical criteria that impact the system's structural integrity. Locations with low wave heights (1-2m or below) situated outside predictable tropical cyclone paths provide the most suitable conditions for implementing the SeaOasis.

Figures 3.11 and 3.12 show the best and good conditions of locations with suitable year-round (best) or seasonally (good) high constant Sea Surface Temperature and Photosynthetically Active Radiation because highly efficient and reliable yields have been observed in laboratory conditions under optimal constant light and temperature levels.

Best conditions - constant year round (in E/m²day)   35 - 45   45 - 55   55 - 65

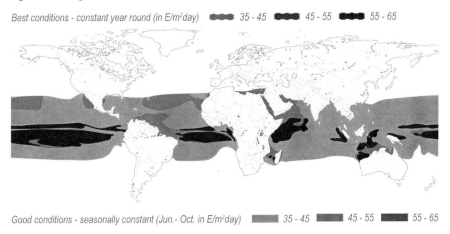

Good conditions - seasonally constant (Jun.- Oct. in E/m²day)   35 - 45   45 - 55   55 - 65

Fig. 3.12 Mean photosynthetically active radiation [9]

Fig. 3.13 Best and very good environmental conditions

Figure 3.13 overlays the above environmental parameters, highlighting specific good and best condition areas with optimal temperature, solar radiation, and wave height. The map was subsequently used to identify zones with optimum environmental conditions for the proposed species of each configuration.

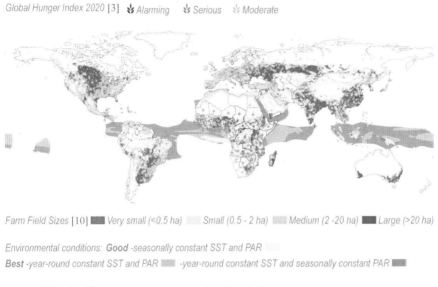

Fig. 3.14 High food insecurity and prevalence of smallholder farms

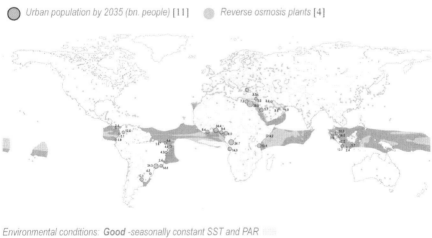

Fig. 3.15 Large predicted urban population growth and desalination plants for SeaOasis configurations A–E

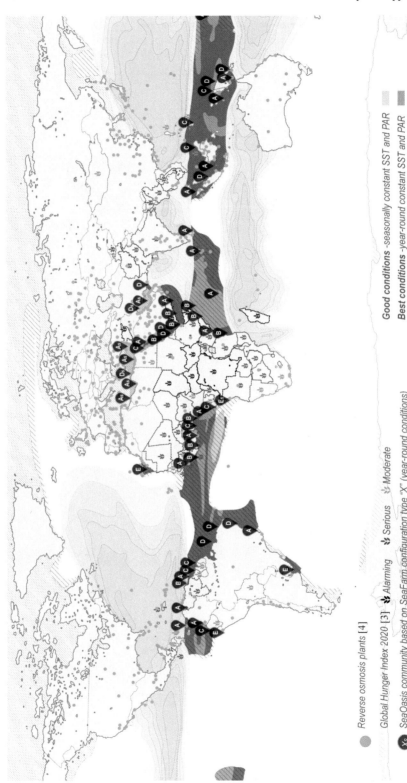

Fig. 3.16 Best and very good conditions for smallholders' SeaOases configurations A–E

Figures 3.14 and 3.15 identify levels of food security and prevalence of small farms as well as providing an indication of future population growth and locations of large desalination plants that could provide brine reject for the farms. These factors relate to the feasibility of implementation and have been used to select SeaOasis configurations with appropriately tailored cost and risk for each implementation region (e.g. likely investment capital availability, locally available materials and technology etc.).

Conclusively, Figure 3.16 identifies the best zones for the SeaOases' implementation meeting the combined criteria for best or very good conditions for all of the above parameters.

References

1. UN World Food Programme (2022) Smallholder market support. https://www.wfp.org/smallh older-market-support. Accessed 05 Jan 2022
2. Averaged national farm sizes, yearly incomes and household sizes based on FAO (2013) Small-holder Data Portrait. www.fao.org/family-farming/data-sources/dataportrait/indicator-details/en/?ind=83447. Accessed 05 Jan 2022
3. Based on Grebmer K, Bernstein J, Alders R et al (2020) 2020 global hunger index: one decade to zero hunger: linking health and sustainable food systems. https://www.globalhungerindex.org/pdf/en/2020.pdf. Accessed 05 Jan 2022
4. Map based on Villacorte L, Tabatabai S, Dhakal N, Amy G, Schippers J, Kennedy M (2014) Algal blooms: an emerging threat to seawater reverse osmosis desalination. Desalin Water Treat 55:2601–2611. https://doi.org/10.1080/19443994.2014.940649
5. Tredici MR, Rodolfi L, Biondi N, Bassi N, Sampietro G (2016) Techno-economic analysis of microalgal biomass production in a 1-ha green wall panel (GWP®) plant. Algal Res (Amsterdam) 19:253–263. https://doi.org/10.1016/j.algal.2016.09.005
6. Map based on MunichRE (2011) Nathan world map of natural hazards. https://catalogue.unccd.int/Map_NATHAN%20-%20World%20map%20of%20natural%20hazards.pdf. Accessed 05 Jan 2022
7. Map based on Meteorological Service of New Zealand (2022) Hindcast maps. https://app.met oceanview.com/hindcast/. Accessed 05 Jan 2022
8. Map based on Giovanni generated seasonal average sea surface temperature maps (MODISA_L3m_SST_Monthly_4km_vR2019.0 data set) compiling data over the past 10 years (2010–2020). Seasonal maps have been overlaid to identify year-round and seasonal temperature zones. https://giovanni.gsfc.nasa.gov/giovanni/. Accessed 05 Jan 2022
9. Map based on Giovanni generated seasonal average photosynthetically active radiation maps (MODISA_L3m_PAR_v2018 dataset) compiling data over the past 10 years (2010–2020). Seasonal maps have been overlaid to identify year-round and seasonal PAR zones. https://gio vanni.gsfc.nasa.gov/giovanni/. Accessed 05 Jan 2022
10. European Commission Joint Research Center, Smallholder Agriculture—Global distribution of farm field size map. https://wad.jrc.ec.europa.eu/smallholderagriculture. Accessed 05 Jan 2022
11. Based on UN, Smith DA, CASA, UCL (2018) World city populations 1950–2035. https://lum inocity3d.org/WorldCity/#/4/36.24/110.83. Accessed Jan 05 2022

Conclusion

This research and design project developed an aqua-farming system called SeaOasis with the aim to tackle the substantial problem of global food supply and security. In terms of creating accessibility and incentives for smallholder families, the proposal presents several advantages compared to traditional land-based farming systems: The small footprint of the SeaOasis allows for easier access as well as decreased maintenance and harvest labour difficulty and time while providing high yearly revenues: Depending on the region, yearly revenue per family could increase dramatically.

While there is great potential for further growth of the aquaculture sector, the intensification of production implied is challenging due to significant potential environmental impacts. New sustainable aquaculture development strategies such as the (nearly) closed-loop system used in the SeaOasis are, therefore, needed to mitigate the environmental impacts of increased aquaculture production.

Such strategies need to harness further technical developments in, for example, feeds, genetic selection, bio-security and disease control, and digital innovation, with business developments in investment and trade. The priority should be to develop aquaculture in regions where population growth will challenge food systems most.

The interlinked approach provided opportunities to:

- expand global crop areas by utilizing the ocean surface for crop production
- address food security by tailoring solutions that support smallholder farming
- optimize food production, resource utilization and economic income by tapping into emerging markets that utilize highly efficient, marine adapted species (e.g. micro-algae) and generate high-value products
- make emerging market technologies available to smallholder farmers in order to provide access to global markets and incentives for investment in smallholdings
- utilize sustainable farming practices typically employed by smallholders.

J. Baumeister and I. C. Giurgiu, *SeaOasis*, SpringerBriefs in Architectural Design and Technology, https://doi.org/10.1007/978-981-19-1373-0

While the potential benefits of such an approach are numerous, we recognize that creating practical applications that work on multiple, interlinked levels is not without challenges. Innovative approaches are required, especially for the aspect of adapting/merging traditional sustainable agriculture practices to new materials, technologies, and markets.

Given the herewith proven capability of the SeaOasis concept to meet future food demand and to increase at the same time smallholders' income, practical applications as prototypes should be tested as a next step. The required little financial input appears more than justified in the light of the achievable revenues and global food security.

Appendix A

See Table A.1.

J. Baumeister and I. C. Giurgiu, *SeaOasis*, SpringerBriefs in Architectural Design and Technology, https://doi.org/10.1007/978-981-19-1373-0

Table A.1 Optimal growth parameters for potential cultured species

Type	Species	Application	Temp. (°C)	Light	
				Intensity (lux)	Photoperiod (light : dark h)
Freshwater					
	Nile Tilapia	food	31-36	-	-
	Ictalurus punctatus	food	24-30[29]	-	-
Saltwater					
	Oreochromis mossambicus	Food	25-30[22]	-	-
	Sparus Aurata (gilthead seabream)	Food	26-30 (25 optimal) [23]	-	-
	Sciaenops ocellatus (red drum)	Food	26-30 (28 optimal)[25]	-	-
Saltwater					
	Crassostrea gigas	Food	20-30[35]	-	-
	Ostrea edulis	Food	25[37]	-	-
	Mytilus galloprovincialis	Food	13-25[36]	-	-
Freshwater					
	Corbicula fluminea (Asian clam)	Food	2-30[38]	-	-

FISH

BIVALVE

Salinity (g/l)	pH	Yield (kg/m^{-2}yr^{-1})	Market price (USD/kg)	Notes:
1[31]	6.5-9	0.65[32]	4[28]	
<4[29]	-	0.18[30]	1.26[30]*	*2003 price
32-40[22]	6-8[22]	2.1[26]*	3.6[34]	*yield based on Thailand pond case
18-28[24]*	-	28[11]	4.45[28]**	*optimum for fry growth but adults adapted to higher range of salinity (brackish to seawater) ** minimum price considered
32-36[25]	8[25]	0.42[27]	6.61[33]*	*2016 price
20-30[35]	7-9[35]	10-15[42] kg/ m rope	19.69[28]	
>20[37]	-	10-15[42] kg/ m rope	16.45[28]	
27-40[36]	-	8[40]-11[41] kg/ m rope	2.25[28]	
<13[38]	-	no data*	3[39]	* Subsequent calculations assume 10 kg/m rope as per similar species

Table A.1 (continued)

Type	Species	Application	Temp. (°C)	Light	
				Intensity (lux)	Photoperiod (light : dark h)
Micro					
	Chlorella vulgaris	Food, Aquaculture fish feed (suplement); biofuels (anaerobic digestion)	24[1]	12960	12:12[2]
	Arthrospira platensis (Spirulina)	Food, Aquaculture fish feed (suplement); biofuels	30-35[3]	1500-2500 [3] (450nm wavelength); 2000 (white flourescent) [4]	12:12[4]
	Dunaliella salina	Food and animal feed (β-carotene extracts)	20[5]	18000[5]	12:12[6]
	Nannochloropsis sp.	Biofuel more frequently	24-26[7]	5920[7]; 4000[8]	not specified
Macro					
	Ulva lactuca	Fish feed	18.2-31.2[11]	-	-
	Eucheuma spp	Carrageenan for food and cosmetics	27-30[14]	27000-48600[14]*	-
	Gracilaria lichenoides	Agar,paper,biofuel[15]; Filtration and feed in fish/shrip tanks [16]	30[12]; 31.3[13]	15510[13]	-

ALGAE

Salinity (g/l)	pH	Yield (kg/m^{-2}yr^{-1})	Market price (USD/kg)	Notes:
0-15 ; 38.5[2]	7-8	6.97[1]	44[10]	Note salinity based on [2] is for a saltwater tolerant strain of C. vulgaris ; [1] see formula to lux from mmol/m2s-1 based on https://www.apogeeinstruments.com/conversion-ppfd-to-lux/ - sunlight assumed
20-70	8.5-10.5	~2.19[3]	42[10]	
45[5]	8[6]	~9.12*;	1500 for pure β-carotene [19]**	[5] see formula to lux from mmol/m2s-1 based on https://www.apogeeinstruments.com/conversion-ppfd-to-lux/ cool white flourescent assumed; * yield approx. based on [6] productivity of 2.8 g/l over 11 days; **note approx 14% b-carotene from total dry yield [19]
20-40[7]	7.4-7.8[7]	~8.76*	100[20]	* based on [9] general value and averaged likeliest yield scenario
41[11]	8.5-8.9[11]	78[11]	-	no price shown as proposed use is for fish feed within closed loop system
30-35[14]	6.8-8.2[18]	36[17]	0.55 dried raw[14]**	*see formula to lux from mmol/m2s-1 based on https://www.apogeeinstruments.com/conversion-ppfd-to-lux/ w sunlight assumed; **Philippines 2014 raw material price
35[12]	7-8[13]	12.7[15]**	18[15]**	*see formula to lux from mmol/m2s-1 based on https://www.apogeeinstruments.com/conversion-ppfd-to-lux/ - sunlight assumed; ** [15]approx. tank growth yields considered and price for agar in 2009

References

1. Amini H, Wang L, Shahbazi A (2016) Effects of harvesting cell density, medium depth and environmental factors on biomass and lipid productivities of Chlorella vulgaris grown in swine wastewater. Chem Eng Sci 152:403–412. https://doi.org/10.1016/j.ces.2016.06.025
2. Luangpipat T, Chisti Y (2017) Biomass and oil production by Chlorella vulgaris and four other microalgae—effects of salinity and other factors. J Biotechnol 257:47–57. https://doi.org/10.1016/j.jbiotec.2016.11.029
3. Soni R, Sudhakar K, Rana R (2019) Comparative study on the growth performance of Spirulina platensis on modifying culture media. Energy Rep 5:327–336. https://doi.org/10.1016/j.egyr.2019.02.009
4. Kumar M, Kulshreshtha J, Singh G (2011) Growth and biopigment accumulation of cyanobacterium Spirulina platensis at different light intensities and temperature. Braz J Microbiol 42:1128–1135. https://doi.org/10.1590/S1517-83822011000300034
5. Abu-Rezq T, Al-Hooti SN, Jacob DA (2010) Optimum culture conditions required for the locally isolated Dunaliella salina. J Algal Biomass Utln 1(2):12–19
6. Kim W, Park J, Gim G et al (2011) Optimization of culture conditions and comparison of biomass productivity of three green algae. Bioprocess Biosyst Eng 35:19–27. https://doi.org/10.1007/s00449-011-0612-1
7. Abu-Rezq TS, Al-Musallam L, Al-Shimmari J et al (1999) Optimum production conditions for different high-quality marine algae. Hydrobiologia 403:97–107. https://doi.org/10.1023/A:1003725626504
8. Imamoglu E, Demirel Z, Conk Dalay M (2015) Process optimization and modeling for the cultivation of Nannochloropsis sp. and Tetraselmis striata via response surface methodology. J Phycol 51:442–453. https://doi.org/10.1111/jpy.12286
9. Mu D, Min M, Krohn B et al (2014) Life cycle environmental impacts of wastewater-based algal biofuels. Environ Sci Technol 48:11696–11704. https://doi.org/10.1021/es5027689
10. Barkia I, Saari N, Manning S (2019) Microalgae for high-value products towards human health and nutrition. Marine Drugs 17:304. https://doi.org/10.3390/md17050304
11. Neori A, Shpigel M, Ben-Ezra D (2000) A sustainable integrated system for culture of fish, seaweed and abalone. Aquaculture 186(3–4):279–291. https://doi.org/10.1016/S0044-8486(99)00378-6
12. Raikar S, Iima M, Fujita Y (2001) Effect of temperature, salinity and light intensity on the growth of Gracilaria spp. (Gracilariales, Rhodophyta) from Japan, Malaysia and India. Indian J Mar Sci 30:98–104
13. Xu Y, Wei W, Fang J (2009) Effects of salinity, light and temperature on growth rates of two species of Gracilaria (Rhodophyta). Chin J Oceanol Limnol 27:350–355. https://doi.org/10.1007/s00343-009-9116-0
14. Gavino C, Trono Jr (2022) Eucheuma spp. In: FAO Fisheries and Aquaculture Division. Cultured Aquatic Species Information Programme. https://www.fao.org/fishery/en/cultureds pecies/eucheuma_spp/en. Accessed 19 Jan 2022
15. Santelices B (2022) Gracilaria spp. In: FAO Fisheries and Aquaculture Division. Cultured Aquatic Species Information Programme. https://www.fao.org/fishery/en/culturedspecies/gracilaria_spp/en. Accessed 19 Jan 2022
16. Xu Y, Fang J, Wei W (2008) Application of Gracilaria lichenoides (Rhodophyta) for alleviating excess nutrients in aquaculture. J Appl Phycol 20:199–203. https://doi.org/10.1007/s10811-007-9219-y
17. FAO (1990) Site selection for Eucheuma spp. farming. http://www.fao.org/3/AB738E/AB7 38E03.htm#ch3. Accessed 19 Jan 2022
18. Rupidara A (2019) The growth of Eucheuma spinosum in different depth of Batubao marine waters West Kupang. Ecol Environ Conserv J 25(Aug Suppl Iss):162–167
19. Paniagua-Michel J (2015) Microalgal nutraceuticals. In: Kim S (ed) Handbook of marine microalgae: biotechnology advances. Academic Press, Amsterdam NL, p 255

20. Zghaibi N, Omar R, Kamal S et al (2019) Microwave-assisted brine extraction for enhance-
     ment of the quantity and quality of lipid production from microalgae Nannochloropsis sp.
     Molecules 24(19):3581. https://doi.org/10.3390/molecules24193581

21. Alceste C (2017) Considerations for tilapia farming in saltwater environments. https://www.
     aquaculturealliance.org/advocate/considerations-tilapia-farming-saltwater-environments/.
     Accessed 19 Jan 2022

22. Fitzsimmons K (2022) Oreochromis mossambicus (Mozambique tilapia). In: Invasive species
     compendium. CAB International, Wallingford. https://www.cabi.org/isc/datasheet/72085#
     toclimate. Accessed 19 Jan 2022

23. Heather F, Childs D, Darnaude A, Blanchard J (2018) Using an integral projection model
     to assess the effect of temperature on the growth of gilthead seabream Sparus aurata. PLOS
     ONE 13:e0196092. https://doi.org/10.1371/journal.pone.0196092

24. Volstorf J (2021) Sparus aurata (Findings). In: FishEthoBase (ed) Fish ethology and welfare
     group. http://fishethobase.net/db/49/findings/. Accessed 19 Jan 2022

25. Treece G, Adami R (2022) Sciaenops ocellatus (red drum) In: Invasive species
     compendium. CAB International, Wallingford. https://www.cabi.org/isc/datasheet/64720#
     toclimate. Accessed 19 Jan 2022

26. Rakocy JE (2022) Oreochromis niloticus. In: FAO Fisheries and Aquaculture Division.
     Cultured Aquatic Species Information Programme

27. Lutz CG, Wolters WR, Landry WJ (1997) Red drum Sciaenops ocellatus field trials: economic
     implications. J World Aquacult Soc 28:412–419. https://doi.org/10.1111/j.1749-7345.1997.
     tb00289.x

28. FAO/GLOBEFISH (2020) European fish price report. https://issuu.com/globefish/docs/
     epr__may_2020. Accessed 19 Jan 2022

29. FAO (2022) Channel catfish—Ictalurus punctatus. http://www.fao.org/fishery/affris/species-
     profiles/channel-catfish/channel-catfish-home/en/. Accessed 19 Jan 2022

30. Stickney RR (2022) Ictalurus punctatus. In: FAO Fisheries and Aquaculture Division.
     Cultured Aquatic Species Information Programme. https://www.fao.org/fishery/en/cultureds
     pecies/ictalurus_punctatus/en. Accessed 19 Jan 2022

31. Lawson EO, Anetekhai MA (2011) Salinity tolerance and preference of hatchery reared Nile
     Tilapia, Oreochromis niloticus (Linneaus 1758). Asian J Agric Sci 3(2):104–110

32. Middendorp AJ (1995) Pond farming of Nile tilapia, Oreochromis niloticus (L.), in northern
     Cameroon. Mixed culture of large tilapia (>200 g) with cattle manure and cottonseed cake
     as pond inputs, and African catfish, Clarias gariepinus (Burchell), as police-fish. Aquac Res
     26:723–730

33. Monterey Bay Aquarium Seafood Watch (2016) Red drum Sciaenops ocellatus. https://sea
     food.ocean.org/wp-content/uploads/2016/10/MBA_SeafoodWatch_RedDrumReport.pdf.
     Accessed 19 Jan 2022

34. Salia AMJ (2008) Economic analysis of small-scale Tilapia aquaculture in Mozambique.
     https://www.grocentre.is/static/gro/publication/11/document/alda08prfa.pdf. Accessed 19
     Jan 2022

35. Goulletquer P, Bonham V (2022) Magallana gigas (Pacific oyster) In: Invasive species
     compendium. CAB International, Wallingford. https://www.cabi.org/isc/datasheet/87296#
     towaterTolerances. Accessed 19 Jan 2022

36. Bonham V, Riginos C, Shields J (2022) Mytilus galloprovincialis (Mediterranean mussel).
     In: Invasive species compendium. CAB International, Wallingford. https://www.cabi.org/isc/
     datasheet/73756#towaterTolerances. Accessed 19 Jan 2022

37. Goulletquer P (2022) Ostrea edulis. In: FAO Fisheries and Aquaculture Division. Cultured
     Aquatic Species Information Programme. https://www.fao.org/fishery/en/culturedspecies/
     ostrea_edulis/en. Accessed 19 Jan 2022

38. Allen U (2022) Corbicula fluminea (Asian clam). In: Invasive Species Compendium. CAB
     International,   Wallingford.https://www.cabi.org/isc/datasheet/88200#towaterTolerances.
     Accessed 19 Jan 2022

39. Lee S, Azree A, Ramli M et al (2018) Growth and response of Asian clam, Corbicula fluminea, towards treated quail dung. Int J Aquat Sci 9(2):120–121

40. Karayücel S, Celik MY, Karayücel I, Erik G (2010) Growth and production of raft cultivated Mediterranean Mussel (Mytilus galloprovincialis Lamarck, 1819) in Sinop, Black Sea. Turk J Fish Aquat Sci 10:09–17. https://doi.org/10.4194/trjfas.2010.0102

41. Buck B (2007) Experimental trials on the feasibility of offshore seed production of the mussel Mytilus edulis in the German Bight: installation, technical requirements and environmental conditions. Helgol Mar Res 61:87–101. https://doi.org/10.1007/s10152-006-0056-1

42. Danioux C, Bompais X, Paquotte P, Loste C (2000) Offshore mollusc production in the Mediterranean basin. In: Muir J, Basurco B (eds) Mediterranean offshore mariculture. CIHEAM, Zaragoza, pp 115–140. https://om.ciheam.org/om/pdf/b30/00600655.pdf. Accessed 19 Jan 2022

Appendix B

Table B.1 Indicative cost and revenue estimate for each unit type

Total costs:

Base Unit: **$11,540**

| *Filter/ Hydroponic units:* | circular **$23,880**; spiral **$20,517**; radial **$25,610** |
| *Algae/ Mixing units:* | circular **$24,590**; spiral **$21,073**; radial **$26,226** |

Full 5-unit loop: from ca. **$99,500** to **$113,000** depending on unit types used

Species		Chlorella vulgaris	Arthrospira platensis	Dunaliella salina	Nannochloropsis sp.
	Algae	Chlorella vulgaris	Arthrospira platensis	Dunaliella salina	Nannochloropsis sp.
	Oyster	Crassostrea gigas	Ostrea edulis	Mytilus galloprovincialis	Corbicula fluminea
Revenues (USD/yr)					
Configuration	Circular	$18,401	$5,519	$114,912	$52,560
		$47,610	$39,776	$4,185	$5,580
	Spiral	$14,414	$4,323	$90,014	$41,172
		$33,020	$27,587	$2,903	$3,870
	Radial	$15,947	$4,783	$99,590	$45,552
		$41,467	$34,644	$3,645	$4,860

Note :
1. All prices in USD and calculated at 2020 online US or Australian listings
2. Prices exclude welding, shipping, pumps, generators
3. Revenues shown respresent gross totals calculated on the basis of pricing and yields shown in Appendix A and Table B.2 quantities.
4. Due to decreased PBR areas, algae revenues for Mixing units should consider a third of the Fish/Algae unit revenue.

© The Author(s), under exclusive license to Springer Nature Singapore Pte Ltd. 2023
J. Baumeister and I. C. Giurgiu, *SeaOasis*, SpringerBriefs in Architectural Design
and Technology, https://doi.org/10.1007/978-981-19-1373-0

Table B.2 Indicative quantities and cost breakdown for each unit type

		Item	Unit	Quantity	Cost/unit	Cost	Assumptions
PBR and Float systems	Circular option	Algae bags	m2	60	11.84	710.4	~ 54 1x1m bags @ 1.5m centres
		Rope seaweed	m	80	7.8	624	Assumed 12mm jute rope
		Rope oyster	m	186	7.8	1450.8	
		Marker buoy	pcs	54	11.49	620.46	Assumes 4.5"x16" fenders
		Flexible tubing	m	136	31.21	4244.56	
		Algae bag connectors	pcs	54	100	5400	provisional sum for SLS printing
	Spiral option	Algae bags	m2	47	11.84	556.48	~ 40 1x1m bags @ 1.5m centres
		Rope seaweed	m	66	7.8	514.8	Assumed 12mm jute rope
		Rope oyster	m	129	7.8	1006.2	
		Marker buoy	pcs	40	11.49	459.6	Assumes 4.5"x16"
		Flexible tubing	m	96	31.21	2996.16	
		Algae bag connectors	pcs	40	100	4000	provisional sum for SLS printing
	Radial option	Algae bags	m2	52	11.84	615.68	combined 1x0.6m and 1x1m panels
		Rope seaweed	m	68	7.8	530.4	Assumed 12mm jute rope
		Rope oyster	m	162	7.8	1263.6	
		Marker buoy	pcs	63	11.49	723.87	Assumes 4.5"x16"
		Flexible tubing	m	101	31.21	3152.21	
		Algae bag connectors	pcs	84	100	8400	provisional sum for SLS printing
Base system		Rim connector	pcs	1	800	800	provisional sum for custom profile
		ETFE base	m2	200	20	4000	
		Inner tubes	pcs	79	60	4740	Assumed 12.4R38 tubes
		Tube container membrane	pcs	1	2000	2000	provisional sum for custom design

Note :

1. All prices in USD and calculated at 2020 online US or Australian listings

2. ProviAPT reactor prices as per the maximmum estimate of 10euro/m^2 [1]

3. Connectors assumed SLS rapid prototyping with provisional price 0.96 USD per cm^3 and ca. 100cm^3 volume

Reference

1. Roef L, Jacqmain M, Michiels M (2012) 13 Case study: microalgae production in the self-supported ProviAPT vertical flat-panel photobioreactor system. In: Posten C, Walter C (eds) Microalgal biotechnology: potential and production. De Gruyter, Berlin, Boston, pp 243–246. https://doi.org/10.1515/9783110225020.243

Printed in the United States
by Baker & Taylor Publisher Services